Global Congress on Process Safety 2024 - Abstracts Only

Held at the 2024 AIChE Spring Meeting and 20th Global Congress on Process Safety

New Orleans, Louisiana, USA
24-28 March 2024

ISBN: 979-8-3313-1697-6

Printed from e-media with permission by:

Curran Associates, Inc.
57 Morehouse Lane
Red Hook, NY 12571

Some format issues inherent in the e-media version may also appear in this print version.

Printed with permission by Curran Associates, Inc. (2025)

For permission requests, please contact AIChE
at the address below.

AIChE
120 Wall Street, FL 23
New York, NY 10005-4020

Phone: (800) 242-4363
Fax: (203) 775-5177

www.aiche.org

Additional copies of this publication are available from:

Curran Associates, Inc.
57 Morehouse Lane
Red Hook, NY 12571 USA
Phone: 845-758-0400
Fax: 845-758-2633
Email: curran@proceedings.com
Web: www.proceedings.com

TABLE OF CONTENTS

Effectiveness of Safety Tools and Management

Salvador Avila Filho, Ednêi Couto de Lima Júnior Sr., Isabel Leite Miralha Alves

The Variability of factors that affect human performance can cause process losses with severity that affects: Reliability; Occupational Safety, Operational, Process; and Public Safety. All these impacts on the neighboring community. The investigation into the nature of this variability can indicate new ways of acting in Process Safety and in the review of assessment tools and, barriers to avoid accidents. We introduced the analysis of Accidents through the human factors network that interconnects the operational routine until the system stop and consequences. In terms of human performance, we indicate the relationships between behavioral aspects that affect cognitive quality and authorize, or trigger, chain events until the accident. At that moment, during the debate on deviation normalization, we discussed the importance of the following aspects: Human Behavior, Variability and Compensation; Effectiveness of Security Barriers; Circulation of Information in the Construction of Knowledge during the decision-making process; Lessons learned from the Military Services for the Chemical Industry; and what is the noise source for the mind map of the engineer and operator in the industry. This Research is in progress and have been published in books by Elsevier (2022) and Francis & Taylor (2024) and been sold by more than 25 international sites.

The Effectiveness of safeguard tools for Process Safety depends on their integration with Human Reliability in the Technological environment, both during the design and operation phase. Incorrect conceptual assumptions in establishing these barriers are: (a) The Technical Design is inherently safe; (b) LOPA Protection Layers are independent; and (c) The Operational Discipline privileges the standard considering the classical static risk, working with punishment and positive reinforcement, simplifying prevention actions.

This work intends to discuss the concepts and corrections in the steps and in the application of Process Safety tools. The 1st Layer of LOPA does not consider the reasons for human errors in the project discussion. It only inserts the Engineering ignorance coefficient, increasing the plant's capacity to compensate the reduction in reliability, due to human error. This situation makes it difficult for the work of process optimization and increase the load in an irregular way.

On the other hand, the Sociotechnical Project of industrial installations is inherently Safe. As it considers the integration of social and technical factors to design the defenses that avoid the reduction of reliability and the accident. The LOPA Protection Layers are not independent, making the current Perception of Security unrealistic, bringing a false comfort of efficiency, for example, of 80%, when in fact, we are halfway in relation to Security, 50%. The lack of a multidisciplinary vision in the investigation of routine events also affects the result, bringing a false security with the respective recommendations for the Project Analysis: PRA, PHA, HAZOP, LOPA, Bowtie, Vulnerability, ALARP, FMEA, SIS, and FTA. Thus, the Vision refers only to tangible issues, without discussing culture, communication, and cooperation. It becomes an incomplete object of research in relation to the causes of a probable accident.

On the other hand, the operational discipline tool OD COO, follows the ABC line, reinforcing actions considered correct - operational conduct - making the negative consequences visible and repressing what can cause the accident. Then, the technical, operational, and social complexity is not discussed, reducing the visibility of the intrinsic hazard changes and, hazard action mechanisms changing.

The Mind Map changes, decisions change according to undiscussed aspects such as: social conflicts, changes in priority, and changing motivating factors for work. The discussion of the effectiveness of Process Safety tools will indicate adjustments, calibrations, and necessary changes about a real and non-fictional decrease in the probability of accidents and disasters.

The Challenge: How to Get Success in New Problems with Old Solutions

Alberto Coelho Sr.

It's a fact that society is evolving, but it's also a fact that this evolution is taking place in a wide variety of areas, bringing problems that didn't exist before. Sometimes, it is not easy to think that what brings technological advances can also present difficulties. This scenario could be seen if charged by a battery. We need to pay attention to the value of the battery level because if the watch runs out of battery and the owner is in a place where there isn't a battery to sell, he won't know the time. It's important to remember that if it were a winding model, there wouldn't be this problem. So, with this example, we can see that modernity is not always associated with solutions and that we must resort to old solutions. The purpose of this paper is to show where this resource can be applied in the process plants.

Contained Caution: A New Era in Tank Risk Assessments

Idil Osmanli, Tekin Kunt

The Buncefield explosion in Hertfordshire, UK, underscores the profound implications of overlooked hazards and gaps within tank risk assessment protocols. This catastrophic event, resulting from multiple failures for a fuel storage terminal, led to one of the most significant peacetime fires that Europe has witnessed, yielding extensive economic, environmental, and societal repercussions. The incident demonstrated that the existing risk assessment at Buncefield missed several critical elements, notably a lack of preparation for high-consequence, low-frequency events, inadequate prevention measures against tank overfilling, and insufficient maintenance and reliance on safety instrumented systems. The failure to fully recognize the risks associated with vapor cloud formation and ignition, combined with a complacent attitude towards safety procedures, further expedited the disaster.

This paper entitled "Contained Caution: A New Era in Tank Risk Assessments" presents an evolved and comprehensive approach in response to the shortcomings revealed by the Buncefield incident. This new methodology emphasizes systematic hazard identification, routine and meticulous hardware inspections, strict adherence to operational protocols, and the fostering of an anticipatory safety culture. The proposed framework also endorses the utilization of sophisticated technological systems for constant tank monitoring, comprehensive training regimes that promote acute situational awareness, and the crafting of detailed emergency response strategies.

This approach, labeled "Contained Caution," does not just aim to improve standards, but it also intends to revolutionize them. By broadening the scope of traditional risk assessments and instilling a pervasive, collective responsibility for safety, this novel approach seeks to prevent future calamities, protect communities and ecosystems, and reinforce unwavering commitment to both operational excellence and the benefit of the public.

Vapor Cloud Explosions Involving Separated Congestion Volumes – Application of the BST Method

Ryan Hart, Nicholas Reding, May Yen, Alfonso Ibarreta, Delmar Morrison

Vapor Cloud Explosion (VCE) overpressure calculations for potential flammable vapor release scenarios deliver critical information for the design and construction of contemporary oil & gas processing facilities. Multiple techniques exist for modeling the effects of overpressure hazards, from simplified empirical models to increasingly robust complex computational fluid dynamics (CFD) models. Alternatively, empirical models (e.g., TNT method, TNO method, BST method) function at a lower computational cost and are capable of reasonable overpressure estimates.

The Baker–Strehlow–Tang (BST) method is a commonly used empirical model for overpressure hazards, offering the means to adjust computation for different fuel types, level of congestion, and the size/position of the congested vapor cloud volume(s). However, a challenge when applying the BST method is appropriately defining these parameters to depict complex, real-world configurations of congestion volumes. The BST method allows the user to designate a congestion level but assumes that the volume sees uniform congestion throughout, something atypical for many industry applications. Furthermore, the BST method is not directly capable of quantifying the effects of multiple congested volumes positioned in series with separation. CFD tools, however, can handle such complicated physical geometry scenarios but often at the cost of computational time and expense.

In this paper, we evaluate the validity of assumptions used in our application of the BST method to quantify VCEs within a series of multiple complex congestion volumes that are separated but occupied by a continuous flammable cloud, an extension of previous works evaluating the degree if congestion within a single volume. Empirical overpressure results will then be compared to simulations that were prepared using a commercial CFD package, to determine the accuracy and reliability of BST congestion simplifications.

Risk Based Safeguard Requirements for Processes with a History of Repeated Explosion Incidents

Jerry Lu

Managing reactive chemical operations without sufficient and adequately designed safeguards in place can lead to incidents with grave consequences, which can happen repeatedly. One example is nitrous oxide explosions in bulk liquid handling processes. In the report of an incident that happened in 2016, CSB stated that since 2001 such events have occurred with an average of one explosion every four years. In total, these incidents killed six workers and injured twenty-one other people. Guidelines on general requirements to control these hazards are available in industrial standards, but the process operator must define the detailed requirements for safeguards, based on the specific process and operation conditions. This paper describes a nitrous oxide tanker unloading process risk assessment that includes selection of inherently safer design options based on material hazard properties, as well as quantitative risk assessment. The assessment addresses how to meet risk tolerability criteria for processes with high demand rates, requiring significant risk reduction from instrumented systems, including how to ensure independence between different IPLs. The paper also provides a summary of how such systems perform on actual operating facilities.

Desenvolvimento Comportamental Em Segurança Como Pilar Estratégico Para Impulsionar Uma Transformação Cultural Organizacional

Tatiana Zajdenweber, Hebson Nery, Thais Almeida

Na Vale acreditamos fortemente que todo acidente pode ser evitado. O nosso foco tem sido trabalhar em 4 pilares: Prevenção de lesões e doenças crônicas, Prevenção de fatalidades, Prevenção de acidentes catastróficos e Respeito ao meio ambiente. Como parte da nossa estratégia, fomentamos uma cultura de segurança trabalhando no engajamento do comportamento seguro em todos os nossos empregados, trazendo a liderança como parte essencial no desenvolvimento destes comportamentos.

Obtivemos como resultado da nossa estratégia uma redução de 72% nos eventos de alto potencial (N1[1] e N2[2]) entre 2018 e 2022, entretanto com o intuito de sustentar e potencializar cada vez mais os nossos resultados, colocando em prática um dos nossos principais comportamentos chave "obsessão por segurança e gestão de riscos", identificamos a necessidade de ampliar os nossos esforços através da capacitação da liderança em desenvolvimento comportamental para alavancar os resultados a patamares mais elevados.

Andy Grove[3] afirma que o estabelecimento da confiança com subordinados é uma das principais características que um líder deve dominar, pois tem um alto impacto na motivação da equipe e Dale Carnegie[4] declara que a motivação é a chave para obter excelentes resultados. Alinhados a estes conceitos, bem como ao pilar do PSM "comprometimento com segurança de processos", mais especificamente aos elementos "envolvimento das partes interessadas", "cultura de segurança de processos", "envolvimento dos trabalhadores" e ao modelo de gestão Vale (VPS – Pilar "Liderança"), impulsionamos a transformação cultural da organização através do programa de desenvolvimento comportamental "Liderar com Segurança", iniciado em agosto de 2022.

Neste artigo demostraremos como este programa estratégico, cujo objetivo é a conscientização dos líderes sobre o seu papel na promoção de um ambiente seguro, capacitando-os para criar um âmbito de confiança com seus subordinados vem elevando o nível da cultura de segurança e corroborando para alavancar a eliminação dos SIFs (serious injury and fatality). Neste sentido, foram empreendidos esforços na capacitação dos líderes, que vem desdobrando o conceito "7 Insights into Safety Leadership" pela Krause Bell Group. Com isto entre 2022 e 2023 foram certificados 13 consultores internos (Diretores e Gerentes Gerais) para difusão do conhecimento, e capacitados +900 líderes em 3 países.

Apresentaremos a robusta estrutura para governança, que conta com a participação da alta liderança (CEO/VPs) bem como a estratégia para implantação do programa contemplando a atuação de stakeholders multidisciplinares tais como as equipes de RH e a comunicação.

Como principais legados deste programa observamos uma maior apropriação do líder no que tange seu papel no engajamento do comportamento seguro, um entendimento mais abrangente sobre o processo para tomada de decisão e suas limitações juntamente com uma melhoria na comunicação de líderes com seus subordinados. Resultado este observado pelo aumento de 508% entre 2018 e 2022 na quantidade de reportes de N3[3] sendo realizados ao longo dos últimos anos.

Dynamic Risk Advisor: A Real-Time Risk Analysis Tool for Process Safety Management

Carlos Junior, Alexandre Bonavita, Victor Costa, Alexandre Aguiar, Gabriel Regis

Dynamic Risk Advisor is a risk management tool developed combining two risk assessment methods (Bowtie and LOPA). This tool is a web application that helps managers, supervisors, and crew members to identify vulnerabilities on operational safety barriers. As a web application, Dynamic Risk Advisor aggregates information from different management systems (e.g., IBM MAXIMO, DRAKE) and shows in real-time the process safety barrier health situation. Furthermore, using LOPA technique it is possible to calculate the hazardous scenario tolerability also in real time, once Dynamic Risk Advisor considers the number of active and degraded barriers.

Iris - Democratizing Process Incident Reporting at Vale

Thiego Cugler de Pontes Sr., Juliana Mello, Patricia Fonseca, Yasmin Souza Sr.

Vale, as one of the largest mining companies in the world, understands the importance of maintaining an unwavering commitment to safety. With around 185 thousand people on its staff, the company recognizes that the lives of its employees come first. After the tragic event in Brumadinho, in which a tailings dam collapsed, causing loss of human life and serious damage to the environment, Vale reaffirmed its commitment to safety. The Brumadinho incident was a critical moment that pushed the company to reevaluate its practices and establish a new ambitious goal: eliminate fatalities by 2025.

The quest to improve safety and prevent incidents similar to the Brumadinho tragedy led Vale to recognize the need for an innovative approach. The incident identification and handling process was complex and required improvements to ensure an effective response. It was in this context that the idea for the Iris tool emerged.

The creation of Iris was based on James Reason's 'Swiss cheese' theory. In this theory, each 'slice' of cheese represents an organizational layer that acts as a barrier to prevent incidents. When multiple barriers fail, accidents occur. Therefore, early identification and effective handling of incidents are crucial. Iris was designed to act as an initial barrier in identifying incidents, collecting essential information, such as the location and severity of the incident, at the time of recording. This provides a rich database for analysis and decision making, reducing the need for intervention from technical areas.

One of Iris's central objectives is to democratize and simplify the incident reporting process. The tool was designed to be accessible to all employees, regardless of their level of familiarity with technical terminology. Iris's simplified language and usability allow any employee to report an incident quickly and effectively. This lowers the barrier to entry for reporting and encourages a participatory security culture.

Process safety is a critical concern in industrial operations, and Iris is designed to include the ability to report events related to process safety. Vale's 'P Events' categorization is used to identify and treat incidents related to process safety. This ensures that all safety areas are covered by the tool.

Iris acts as a front-end for SAP Incident Management, enabling effective integration with Vale's incident management system. This ensures that the subsequent steps of investigation, execution of action plans and follow-up are properly managed. The combination of Iris and SAP Incident Management results in a complete and effective solution.

The democratization of incident reporting was accompanied by an extensive change management campaign. Vale took the opportunity to reinforce workplace safety concepts throughout the company. Publicity materials, such as intranet, banners and wall newspapers, were used to disseminate information. In addition, almost 5 thousand employees were trained to help their colleagues adapt to changes.

The name 'Iris' was chosen to reflect a specific part of the human anatomy: the iris of the eyes. The iris is responsible for the focus of vision. This metaphor conveys the message that, with a watchful eye and a commitment to safety, employees can help focus on dangerous situations that may have undesirable outcomes. Iris represents a tool that empowers employees to keep a vigilant eye on security.

The objective of this work is to present this tool that brought innovation in the company's search for safety and operational excellence. By simplifying the incident reporting process, democratizing access to the tool and integrating it with SAP Incident Management, Vale is building a stronger security culture. The message from Iris is clear: all employees play a vital role in keeping safety first. With Vale's continued commitment to security, Iris is a tool that reflects the vision of a safer and more promising future.

Ibram - Manual De Gerenciamento De Segurança De Processos Para Indústria Da Mineração

Raquel Pinto, Eduardo Francisco, Ariadna Moreira, Wagner Silva, Juliana Mello, Claudia Pellegrinelli

As rupturas das barragens de rejeitos de mineração em Brumadinho e Mariana causaram impactos sem precedentes no histórico da indústria da mineração, causando efeitos ambientais, socioeconômicos e na vida e na segurança das pessoas e das comunidades das regiões afetadas. Esses eventos colocaram em xeque a essência das atividades do setor, desencadeando o surgimento de novas exigências do poder público, dos órgãos representativos e da sociedade em relação ao padrão de performance em segurança e sustentabilidade das operações. Concomitantemente, despertou-se a percepção de que a abordagem com foco em Segurança Ocupacional não seria suficiente para responder a esses novos desafios da indústria mineral, uma vez que a resposta aos fatores contribuintes para os eventos sucedidos abrangia outros aspectos técnicos não contemplados por essa disciplina.

Em resposta às demandas da sociedade, o Instituto Brasileiro de Mineração (IBRAM) publicou, em 2019, a "Carta Compromisso perante a Sociedade", que logo em seguida evoluiu e se transformou nos compromissos de ESG da Mineração, assumindo uma série de pactos para uma mineração segura, sustentável e inclusiva, cumprindo o seu papel de vetor para o desenvolvimento, de indutora da transformação tecnológica e de protagonista no incentivo à economia circular. O documento aborda doze temas, entre eles, a Segurança de Processos, para o qual, foram assumidos três compromissos: (1) Elaborar um guia técnico definindo diretrizes e boas práticas visando implementar uma jornada de segurança de processo para mineração; (2) Criar núcleo de excelência em segurança de processo do setor mineral no IBRAM, para compartilhar e desenvolver boas práticas; e (3) Criar relatório anual interno sobre segurança e acidentes de processo por meio de fóruns específicos entre empresas do setor mineral.

Como parte das iniciativas oriundas dos compromissos de ESG da Mineração foi fundado o Comitê de Gerenciamento de Segurança de Processos do IBRAM, que exerce um papel fundamental no fortalecimento da Cultura de Segurança de Processos na mineração. A atuação do comitê tem promovido a troca de conhecimentos entre os representantes das empresas signatárias e sinergias entre iniciativas de instituições de referência, como o ICMM (*International Council on Mining and Metals)* e o CCPS (*Centre for Chemical Process Safety)*, criando cooperações especiais entre esses órgãos e a mineração brasileira. Uma das entregas deste comitê é a criação do Manual de Gerenciamento de Segurança de Processos para a Indústria de Mineração, elaborado pelo 1° subgrupo de trabalho, baseado em um modelo de sistema de gerenciamento especialmente criado para a mineração e que traz na fundamentação técnica os aprendizados de outros setores, como a indústria química, ao mesmo tempo que considera o estágio de evolução em Segurança de Processos do setor minerário, suas características, requisitos técnicos e normativos, dinâmica e necessidades específicas e as suas peculiaridades.

Através do estabelecimento deste referencial básico para o tema, serão fortalecidas iniciativas dos demais grupos de trabalho constituintes do comitê formados para atender cada um dos três compromissos para a implementação da Segurança de Processos no setor mineral, como o Centro de Excelência, responsável pela divulgação de boas práticas, estímulo ao envolvimento das empresas com o tema e difusão da Cultura de Segurança de Processos, exercendo um papel fundamental para que os objetivos do comitê sejam atingidos e entendidos por todo o público alvo. O Centro de

Excelência está dividido em seis importantes elementos, sendo eles: (1) GT Pilares, (2) Guias SEPRO, (3) Boas Práticas SEPRO, (4) Lições Aprendidas SEPRO, (5) Ranking Boas Práticas / Lições Aprendidas, e (6) Prêmio de Excelência SEPRO. Toda essa estrutura ficará disponível em um sistema web para a consulta das empresas signatárias do IBRAM. O Manual também estabelece uma forte conexão com as entregas do Grupo de Trabalho responsável pela criação do Relatório de Segurança de Processos, uma vez que os indicadores contidos nesse relatório serão a base para medir a performance do sistema de Gerenciamento de Segurança de Processos.

O sistema proposto no manual engloba os fundamentos técnicos e metodológicos das principais normas e sistemas de gerenciamento intencionalmente reconhecidos, como a OSHAS 3132 as Diretrizes para Segurança de Processos Baseada em Risco, podendo ser utilizado como complemento para os Princípios de Mineração do ICMM, em particular para o Princípio 4 (Gerenciamento de Risco). O modelo é constituído por 12 elementos agrupados em 5 eixos temáticos, a saber: (1) Tecnologia, (2) Operações, (3) Instalações, Equipamentos e Estruturas, (4) Pessoas e (5) Liderança, com forte ênfase nas sinergias entre cada elemento e em uma visão sistêmica do Gerenciamento de Segurança de Processos nas organizações. O Manual de Gerenciamento de Segurança de Processos para Indústria da Mineração descreve cada um dos doze elementos do sistema proposto com base em seus princípios, suas características básicas e interação com os outros elementos do sistema, utilizando exemplos práticos das operações típicas da mineração, como aspectos geotécnicos (taludes, pilhas, barragens e cavas), estruturas civis, operações em minas subterrâneas e minas a céu aberto, beneficiamento, transporte, filtragem e pelotização trazidos durante as discussões do grupo multidisciplinar de elaboração, composto por especialistas das empresas signatárias do IBRAM.

Através do seu pioneirismo, o Manual de Gerenciamento de Segurança de Processos para Indústria da Mineração apresenta ao setor o Gerenciamento de Segurança de Processos por meio de um modelo técnico, especialmente adaptado e através de linguagem próxima, simples e familiar às equipes envolvidas no planejamento, projeto, manutenção e operação, que pode ser utilizada como referência para implementação do Sistema de Gerenciamento de Segurança de Processos na sua forma integral ou parcial, em sinergia com os processos já existentes nas organizações para melhorar seus resultados.

Implementation of Dynamic Barrier Management in an Oil and Gas Company: Advantages and Challenges

Ana Figueredo, Deyse Camara, Lais de Oliveira Gabriel, Andre Dantas

The operational safety sector has been using different techniques to identify and analyses risk such as What if, PHA and HAZOP studies. However, with the increasing complexity of installations and the necessity to be compliance with regulations, new tools have been using to enhance risk management throughout the lifecycle of these installations.

Since 2019, a dynamic barrier management tool has been developed at a large oil and gas company to continuously monitor barriers associated with loss of containment major scenarios bowtie structure in order to support decisions in prioritizing the treatment of degradations or unavailability. The methodology represents a 'risk picture' of the processes with a clear and user-friendly visualization. It shows the relationship between the hazards, potential adverse consequences, and the factors that could cause harm, creating a clear differentiation between proactive (preventive) and reactive (mitigatory) risk management. This platform provides an interface that allows many databases integration and makes dynamic monitoring possible.

As part of the tool development process, the types of barriers and its objectives were defined as well as its integrity requirements. Based on these requirements, it was established rules to determine the condition of each barrier (available, unavailable, degraded and contingency) using information from different databases.

Currently, the database integration and barriers governance have not yet been completed. The tool has been tested with fewer databases in 11 offshore installations. The paper presents the gains and challenges observed during the testing of the dynamic barrier management tool.

The outcome of this study shows that dynamic barrier management tool helps an easy visualization of degraded barriers, allowing the risk owner to make assertive decisions to re-establish the barriers conditions or, in case of intolerable risk situations, apply operational shutdown. Furthermore, the tool allows the integration of company data such as training, and maintenance records in the same place.

However, the tool has its limitations in identifying deviations that lead to the degradation or unavailability of barriers, which may be related to the lack of systematized data for automatic capture or the difficulty in defining degradation criteria that evaluate the quality of data registered in different databases and the impact of human factors in barrier's integrity. Therefore, one of the biggest challenges is to overcome these limitations, allowing that any weakness in barriers would be detected.

In addition, some organizational factors require further development, such as establishing standardized verification activities, ensuring efficient reporting processes at both corporate and tactical levels and defining requirement for updating the bow tie diagram.

To address the challenges encountered, the establishment of technical forums with multidisciplinary teams and the analysis of various organizational documents and standards were fundamental in this implementation process. These forums facilitate discussions and the synthesis of diverse perspectives, contributing to a comprehensive understanding of the challenge involved.

In summary, the implementation of the dynamic barrier management tool for offshore facilities has demonstrated its efficacy in visualizing and addressing barrier degradation. Despite certain limitations, these findings show the necessity for continuous improvement in risk management practices in the company.

Process Safety Indicators on Mining – Focus on Geotechnical Asset

Juliana Mello, Filipe Costa, Bruno Menegatti Esperandio, Patricia Fonseca, Frederico Bezerra

For many years, large companies in the industrial sector evaluated safety performance focused on occupational aspects, with indicators based on injuries rates. After the occurrence of major industrial accidents, it was observed that most occupational indicators associated with injuries did not necessarily demonstrate an improvement in the prevention of industrial accidents. Therefore, there was a need to monitor the safety performance of industrial processes safety with specific indicators.

Historically, this monitoring has been carried out in the oil & gas and petrochemical industries, which subsidized international references such as the Process Safety Indicators Guide of Center for Chemical Processes Safety (CCPS) and the standard of the American Petroleum Institute (API 754) Process Safety Performance Indicators for Refining and Petrochemical Industries. The mining industry, due to the absence of specific references, uses these materials as guidelines for the implementation of process safety indicators.

After tailings storage facilities break down on Brumadinho, we percept that mineral sector production chain has characteristics that could result in major accident, which consequences as several as P-36 Platform explosion – Brazil (2001), sugar refinery explosion – Georgia (2008), coal mine accident – China (2014), and others.

On this way, Vale`s have been working daily to identify lesson learning related to Brumadinho, drives the cultural transformation on the company, foment the application of continuous lesson in the management system and reduce risk event of different operational nature.

Due to the specificities of Vale's operations in relation to mining and ore processing when compared to chemical and petrochemical industries, the need was identified to adapt to the concept of process safety events and the adoption of a methodology to identify indicators related to safety of operations

As a result of this adaptation, was elaborated a methodology to identify, record an classification events, resulting on indicators related to operational safety, considering – including – criteria to identification events related to geotechnical assets.

Considering CCPS requirements, assumptions were established that allow for the incorporation of the concept into the management system through a specific definition and well-defined criteria for identifying and classification occurrences, as well as making adaptations into the internal procedures, computerized system for recording occurrences and communication plan for the company.

The definition of the concept to identify process safety events for Vale was not limited to considering "loss of primary containment" of hazardous material, which is normally practiced in industries. The concept was expanded to also consider the unplanned and uncontrolled release of hazardous energy whose consequence could result on damage on health, environmental, property and the community – with loss containment material or not. Such considerations become the trigger for identifying these events and transcend the application of the concept beyond industrial process plants in organizations.

More than definitions customizing, the systematic were organized in to important steps. The first refers to flowchart with three strategic questions that, through the key word, conduct the employee to think about and result, or not, an identification the occurrence as process safety event.

It should be noted that Vale already has a robust system for identifying and dealing with accidents of different sizes, with the new facilities incorporated into the global event management standard. Events that do not fit into the concept are treated with the same rigor and importance as process events.

Answering the three questions as "yes" in the flow, there is the second step where a list of criteria is made available for each level of severity, aligned with the CCPS guidelines for each Tier. With the participation of a multidisciplinary team, the description of the criteria represents the breakdown of guidelines for the reality of Vale's operations.

In this context, specific criteria were established in order to identify events related to geotechnical assets, which are structures such as water storage, tailings and sediment containment dams, mining waste storage facilities, piles of materials, pits and underground works with critical risks to operations.

Professionals with expertise in geotechnical and process safety issues worked on the development of criteria such as "Dam or dike failure with total or partial uncontrolled release of reserved material" as a Tier1 severity event.

The parameters established in the CCPS as well as other documents such as the Global Industry Standard on Tailings Management (GISTM), which is structured into 15 principles that strengthen a holistic view on the management of tailings storage structures, also served as a source of reference. Principle 11 Develop an Organizational Culture that Promotes Learning, Communication and Early Recognition of Problems stands out, which encourages the implementation of lessons learned from the process of internal investigations into incidents and relevant external reports.

In addition to the criteria associated with dams, parameters were developed considering ore piles, pits and slopes, further encompassing the possibility of identifying events in geotechnical assets, making it possible to identify their causes as a lesson to prevent similar accidents.

Therefore, the objective of this work is to present Vale's methodology for identifying process safety events for mining, considering the general context for equipment and operational assets as well as the specific approach for geotechnical assets.

By identifying, recording and handling process safety events, Vale aims to transform the company's organizational culture by adding other actions to current practices that can promote the prevention of serious industrial accidents, bringing visibility to leadership. Through work aimed at promoting the integrity of its facilities, Vale is strengthened to reach another level of safety and operational excellence, becoming a safer and more reliable company, a reference in mining.

Ibram - Implementation Trajectory of Process Safety Indicatorsin the Mining Industry

Juliana Mello, Raquel Pinto, Eduardo Francisco, Ariadna Moreira, Claudia Pellegrinelli

The Brazilan Mining Institute is a non-profit private organization, with more than 160 companies associated. All IBRAM actions are directly related to build new perspectives about Brazilian mining, design strategies and be a leader on the transition of the sector to a more productive, sustainable, safety scenario and more responsible with all around them.

In this context, in 2019 IBRAM – by means of the Compromising Latter, that follows the global tendency, changed responsible actions to attend to Environmental Safety and Government (ESG) – assumed with the society an important treat to contribute to the mining industry transformation, engaged different companies to think together in a better future.

The break dams put at risk the mining industry essentially. Otherwise, the mining can – and have to - be a vector to the development, to be an induction to the technological improvement, active contributor to an inclusive and balanced lifestyle, taking care of people and the environment, according to IBRAM.

With the aim of leveraging cultural transformation, initiatives related to operational safety were included with genuine engagement to give society confidence, admiration and respect for the operation of the mining industry.

For IBRAM, hazard identification and risk management are essential conditions to ensure that the operations of any economic activity are at acceptable safety levels or higher than the required standards.

In this context, the Institute - supported by mining companies – designed three important initiative the has an intention to provide specific actions to introduce Process Safety concept in the sector.

1. Contribute to a new framework of standards and laws that will regularly bring the mining of the future;
2. Promote the creation of a Center of Excellence for operational safety and Research and Development (R&D) in the mineral sector to share and develop good practices;
3. Create annual reports on operational safety, through specific forums between companies in the mineral sector, educational institutions and non-governmental bodies.

After all, according to the International Council on Mining & Metals (ICMM), mining is fundamental to almost every facet of our daily lives, from construction to technological innovations that boost the quality of our lives. However, we need to be aware of the balance of benefits and impacts of operations, making them safer and more sustainable.

As a means of promoting this balance, based on these initiatives, companies were engaged in an important discussion, introducing the concept and added value of the topic. Despite the little theoretical knowledge, it is clear to each participant that by adding process safety guidelines to the company's management system, they reinforce the objectives of being profitable companies, respecting people's lives, the community, protecting the environment and the company assets.

As a way of putting into practice the implementation of these initiatives, a working group was created that currently has approximately 18 companies and 30 effective participants distributed in three working subgroups to specifically conduct each initiative, under the coordination of representatives of the companies seeking direct discussions towards a collective result.

Based on the guidelines of the Center for Chemical Process Safety (CCPS), precursors in the study of process safety, meetings are held to share the general concept with the multidisciplinary group, seeking to encourage maturity in the subject and assertive direction in adapting to the industry of mining.

To prepare the annual report, the group met periodically in specific meetings to understand and build the system for identifying process safety events based on the guidelines in the Guide for Selecting Leading and legging indicators and the experiences of mining companies such as Anglo American, Usiminas, Vale, Nexa, Mosaic, Arcelor Mittal, among others.

Through active listening, information was collected about the reality of the operations of each participating company in the search for an understanding of the specificities of the mining industry, thus allowing the elaboration and definition of criteria for identifying and classifying events at different levels. of severity known as Tiers.

After all, everyone knows that monitoring occupational safety indicators alone is not enough to evaluate the safety of operations and, therefore, it is of fundamental importance to expand the vision of all aspects that can contribute to improving processes.

As a result of the work, all guidelines will be promoted and supported through the Process Safety Management PSM Implementation Manual and through the initiatives of the Center of Excellence, which plays a fundamental role in the governance of the implementation of process safety actions so that guidelines are implemented and aligned with the defined objectives and that best practices are followed.

Through 6 categories, the Center of Excellence drives the creation and adoption of new ideas and new technologies to be implemented:

- Work Group pillars
- Process Safety Guidelines
- Process Safety Best Pratices
- Process Safety Leasson Learning
- Process Safety Ecellence Goal award

Another important point of the Center of Excellence is to promote the training of companies, ensuring that members have the necessary skills and knowledge to perform their functions in the PSM area with excellence.

We emphasize that the manual provides relevant information on the direction of implementation of 12 of the 20 elements created by the CCPS, believing that this is the best way for the theme to be implemented in mining companies in conjunction with the existing management system.

Given this entire context, we saw the construction of IBRAM initiative 3 with the methodology for identifying process safety indicators in the mining industry. This product is intended to be a means of evaluating the performance of the management system of companies that implement this concept, enabling the group to share lessons learned with the creation of the annual report.

Process Safety Metrics - Learnings from Evaluating Management of Change Implementation By Performance on Process Safety Events Investigation

Bruno Menegatti Esperandio, Tatiana Zajdenweber, Eduardo Francisco

Vale S.A., a mining company, world leader on iron ore production is implementing Process Safety Management Based in Risk (RBPS) according to CCPS model since 2018. Vale recognizes that RPBS was originally designed for the chemical industry however also recognize that concepts are perfectly applicable to any industry with adjustments to the nature of the operations. Step by step and respecting culture development Vale has chosen 5 RPBS elements in which Management of Change of was on them. This work is intended to show learnings and experiences on developing the Valé´s Corporate standard of Management of Change and the work to develop associated performance metrics in special relating MoC and PSM incidents that had as cause an improper management.

Critérios e Premissas Para Aplicação De HAZOP e LOPA Em Análises Qualitativas e Semiquantitativas De Risco

Alvaro Souza Junior, Sergio Claudio Marsal

Diversas técnicas estão atualmente disponíveis para identificação de perigos e análise de riscos de processo. Visando otimizar a aplicação dessas técnicas nas diferentes fases do ciclo de vida das suas instalações industriais, a Petrobras estabeleceu uma orientação para uso das técnicas conforme a sua finalidade: identificação de perigos, análise qualitativa de riscos, análise de consequências, análise de frequência, avaliação semiquantitativa de riscos e avaliação quantitativa de riscos.

Particularmente para a análise qualitativa de riscos, a Petrobras utiliza a Análise Preliminar de Riscos (APR) e o Estudo de Perigos e Operabilidade (HAZOP) como técnicas complementares para a identificação de perigos. Dependendo da classificação de severidade dos cenários analisados, o HAZOP é complementado pela Análise de Camadas de Proteção (LOPA).

Embora amplamente difundidas e aplicadas mundialmente nas indústrias de processo, essas técnicas requerem alguns cuidados na sua aplicação de modo a aumentar sua efetividade. O objetivo deste trabalho é apresentar os critérios e premissas adotados pela Petrobras na aplicação das técnicas para análise qualitativa e semiquantitativa de riscos, em particular o HAZOP e a LOPA, dentro do processo de gestão de risco de suas instalações industriais.

A Importância Da APR (Análise Preliminar de Riscos) No Processo De Fiscalização Do Sistema De Gestão De Segurança Operacional (SGSO) Da Anp

Marco Carneiro Sr.

Hoje, além do SGSO (Sistema de Gestão de Segurança Operacional) a ANP (Agência Nacional de Petróleo, Gás Natural e Biocombustíveis) possui outros 4 regulamentos técnicos onde o SGSO está presente na seguinte porcentagem:

- RTSGI - Regulamento Técnico do Sistema de Gestão Integrada - 46,3%
- RTDT - Regulamento técnico de Dutos Terrestres - 20,4%
- SGSS - Sistema de Gestão de Sistemas Submarinos - 54,2%
- SGIP - Sistema de Gestão de Integridade de Poços - 54%

Face a importância do SGSO, este trabalho tem como objetivo, apresentar como uma análise de risco utilizando a técnica APR (Análise Preliminar de Risco) deve atender aos itens exigidos nas Práticas de Gestão do SGSO (Sistema de Gestão de Segurança Operacional) da ANP. A APR, atende também a requisitos das Normas Regulamentadoras do Ministério do Trabalho e Emprego do governo brasileiro como também atende a ABNT (Associação Brasileira de Normas Técnicas),

Um exemplo é a elaboração do Plano de Resposta à Emergência que tem como base os cenários críticos analisados na APR, onde além de atender à Prática de Gestão nº 13 – Planejamento e Gerenciamento de Grandes Emergências no Item 13.2.3, deve atender também à NBR-15219 – Plano de Emergência – Requisitos e Procedimentos, à Resolução CONAMA nº 398 PEI – Plano de Emergência Individual (para offshore) e as Normas Regulamentadoras do Ministério do Trabalho e Emprego do governo brasileiro, como a NR-01 – Disposições Gerais e Gerenciamento de Riscos Ocupacionais - Item 1.5.6 – Preparação para emergências e a NR-20 – Segurança e Saúde no Trabalho com Inflamáveis e Combustíveis - Item 20.15 – Plano de Resposta a Emergências da Instalação.

Além disso, com base no que foi analisado na APR, temos o planejamento de simulados de emergência, gestão de elementos críticos de segurança operacional, gestão das recomendações elaboradas no estudo de risco, gestão de barreiras, gestão de mudanças, treinamento e o mais importante a divulgação dos estudos e seus resultados para equipe operacional que lida diariamente com os riscos da unidade operacional e que atende a diversas práticas de gestão do SGSO, como:

- Prática de Gestão nº 1: Cultura de Segurança, compromisso e Responsabilidade Gerencial,
- Prática de Gestão nº 2: Envolvimento do Pessoal e
- Prática de Gestão nº 3: Qualificação, Treinamento e Desempenho da Força de Trabalho

Quando divulgamos os estudos de risco e seus resultados para equipe operacional e de manutenção, estamos mostrando o comprometimento gerencial na divulgação do estudo e seus resultados, de certa forma treinando e envolvendo a equipe nos riscos de processo os quais eles têm contato diariamente.

Caso seja de interesse tenho uma apresentação em Power Poit que mostra os itens citados e outros.

103c PSM Best Practices of Wanhua

Yifeng Chen

As a large global chemical enterprise, Wanhua Chemical Group Co., LTD. takes PSM as the most important part of its safety management system. After years of exploration, Wanhua has developed its own insights and best practices of PSM .In this lecture, Michael Chen, Senior Vice President of Wanhua will share how Wanhua think and what Wanhua has done about PSM.

55n A Digitalized, Cumulative Risk Management Approach to Maintenance Backlog Prioritization on High Hazard Installations

Tayo Olusanya

When an item of equipment on a plant is due for maintenance or if there is a fault, a work order is created either for preventive or corrective action. The work order outlines the activities to be undertaken to ensure ongoing suitability or to restore the equipment to its intended functional state and it includes a date when the work is expected to be completed. If the work is not done before the set completion date the equipment goes into maintenance backlog. Operating an equipment in backlog especially for a prolong period could compromise its safety and integrity, lead to a major accident, and potentially expose personnel and the environment to severe hazardous consequences.

Monitoring and eliminating or, at least reducing equipment in maintenance backlog to bare minimum is the goal of owners of high hazard installations but it remains a daunting task. Maintenance backlog management is also widely acknowledged in the high hazard industry as a challenge while concerted efforts are being made through knowledge sharing events and best practice publications to inform and advise on effective ways to manage equipment in backlog to protect against potential safety and operational risk exposure.

Organisations often expend considerable time and effort to reduce maintenance backlog and would generally prioritise Safety and Environmental Critical Equipment (SECE) based on company policies and/ or regulatory requirements. However, even with best endeavours the available resources may be too limited to address all critical equipment items in backlog leaving companies having to make the difficult decisions on which critical equipment to prioritise to sustain an acceptable level of safety while in operation. More pertinent are the questions of what criteria should be used for prioritisation, how would it be applied, and what impact does equipment left in backlog have on the installations cumulative risk exposure?

The process of prioritising equipment in backlog is typically done manually. A team of maintenance, safety and integrity personnel reviews all equipment in backlog periodically using their experience, knowledge, and judgment to assess parameters such as function and condition of each equipment, effect of failure on safety and operations, manhours required to clear the backlog etc. This approach can be time consuming, lead to inconsistencies, human errors such as underestimating the importance of a critical equipment, lack of understanding of the effect of the cumulative effect of multiple SECEs in backlog on major risk exposure etc.

This paper presents an alternative approach to prioritising equipment in backlog with focus on SECEs. Underpinned by the principle of cumulative risk management and the barrier management concept, the methodology entails assessment of the purpose and functionality of each SECE in backlog, simulate and rank the impact of different combinations of SECE failures and present the result in order of installation's vulnerability to major risk exposure. The output from this approach provides owners of high-hazard installations immediate vital insight on equipment items whose failure poses the biggest risk exposure, and a robust basis for prioritising and effectively reducing equipment in maintenance backlog.

Evolución Proceso Mdc En Upstream

Luis Kirs Sr.

Evolución Proceso Manejo Del Cambio en Instalaciones del Upstream

El presente trabajo se desarrolla en el Negocio Chubut-Tierra del Fuego, de la Gerencia Convencional Upstream, donde se había implementado formalmente la gestión de MDC: Manejo del Cambio, desde el sector de Ingeniería de Construcciones a partir del año 2014.

Desde su implementación inicial se habían generado desde la operación un total de 167 solicitudes de cambio en instalaciones físicas, de las cuales únicamente 92 continuaron con el proceso MDC. Esto representa que un 45% de las solicitudes fueron rechazadas, canceladas o pasaron a ser parte de un proyecto.

Además, desde la Sala de Control se gestionaba el requerimiento de inhibición de protecciones, los cuales también podían derivar en cambios físicos o de operación en la instalación con la posibilidad de detectarse en un estado tardío.

Con la necesidad de establecer un orden de prioridades para las solicitudes de MDC, revisar el origen del cambio y optimizar los recursos necesarios, es que a partir del año 2023 se toma la decisión de que el proceso sea gestionado desde el sector de Ingeniería de Procesos.

En el trabajo se describe y analizan los principales desafíos encontrados para la correcta implementación del proceso de gestión del cambio: gestión de solicitudes desde la Operación, orden de prioridades, gestión de los recursos necesarios para ejecución, necesidad de revisores especializados, gestión de cambios por fuera del proceso MDC, implementación correcta del procedimiento e importancia de la información de seguridad de procesos (documentación clave de ingeniería).

También se describe cómo se debía adecuar los procesos vigentes para su correcta implementación, asegurando que los cambios que se realicen no introduzcan inadvertidamente nuevos peligros o sin saberlo, aumenten el riesgo de las amenazas existentes.

163a Lecciones Aprendidas En La Implementación De PSM: Priorizando Adherencia Bajo Un Enfoque Centrado En Elementos Comunes – Sgcec.

Felipe Henao

En un entorno industrial de creciente complejidad y competitividad, la seguridad de procesos emerge como una prioridad incontestable. La implantación de un sólido marco de gestión de seguridad de procesos se revela como un proceso esencial para resguardar la integridad de las operaciones y asegurar la protección de las personas y el medio ambiente. Sin embargo, en nuestra búsqueda constante por la excelencia en seguridad, las presiones de terceras partes, a menudo nos vemos tentados a buscar atajos en el camino en pos de resultados más veloces.

En este contexto, basado en mi experiencia liderando e impulsando procesos de implementación en diversas empresas líderes de la industria en LATAM, me propongo compartir las valiosas lecciones extraídas de dichas implementaciones. Este trabajo destaca un enfoque distintivo que, en lugar de buscar atajos, pone en relieve la importancia de priorizar aquellos elementos que suelen socavar la adherencia de procesos, un enfoque dinámico y denominado por el autor 'Validación y Valoración de Elementos Comunes' (VVEC) que se integra con el sistema de PSM (independiente del acervo de conocimiento elegido CCPS, OSHA, regulación local, etc.). En contraposición al enfoque típico de 'Deming' (PDCA), propongo una alternativa novedosa que aborda áreas críticas y establece un mecanismo concreto para lograr resultados más robustos.

Este enfoque se basa en los siguientes pasos:

1. **Selección de Elementos Comunes**: Identificar los elementos comunes en la organización que son críticos para la seguridad de procesos. Estos elementos incluyen, pero no se limitan a: procedimientos estándares, políticas de seguridad, sistemas de entrenamiento, manuales de gestión, entre otros. En este punto una lección aprendida clave es la consolidación del equipo implementador, donde la experiencia, el conocimiento confluyan con un excepcional sentido común y de liderazgo.
2. **Validación de Elementos Comunes**: Evaluar la efectividad y cumplimiento de estos elementos comunes en la organización. Este paso es clave para evitar pérdidas de tiempo innecesarias, enfocando los esfuerzos en validar la adherencia de los procesos subyacentes bajo esquemas de auditoría, revisiones de documentación y encuestas a empleados para asegurarse de que estos se estén siguiendo adecuadamente.
3. **Evaluación de Brechas**: Identificar las brechas entre los estándares de seguridad deseados y la implementación real de los elementos comunes. Esto ayuda a destacar las áreas donde se necesita mejorar y con quizás menor esfuerzo-recurso se pueden obtener mejores resultados. Aquí un valor importante es acondicionar la implementación al contexto de la organización: nacional, multinacional, publica, privada y las diferentes complejidades que manejan sus activos, un enfoque flexible asegura una transición más exitosa.
4. **Acondicionamiento del Proceso**: Tomar medidas concretas para abordar las brechas identificadas. Esto podría implicar la revisión y actualización de documentos, políticas y procedimientos, la implementación de nuevos sistemas de entrenamiento o la asignación de recursos adicionales. En este punto es importante priorizar el trabajo sobre aquellos elementos que cuentan con habilitador tecnológico 'in place' o en proceso de implementación, igualmente aquellos que pudieran obtenerlo mediante la utilización de los medios propios de la organización como: Sharepoint, Bases de Datos, Sistemas expertos como ERP, EAM, entre otros.

Para llevar a cabo esta evaluación de manera efectiva, se propone una herramienta de evaluación específica, denominada como el 'Índice de Cumplimiento de Elementos Comunes en Seguridad de Procesos' (ICEC-SP). Esta herramienta sé compone de un conjunto de indicadores que miden la efectividad de los elementos comunes clave y su alineación con los estándares de PSM. Los indicadores incorporan métricas cuantitativas, como por ejemplo la cantidad de procesos implementados para un elemento del marco de gestión, la cobertura de los habilitadores tecnológicos en dichos procesos, el porcentaje de empleados capacitados o el número de incidentes reportados relacionados con un elemento común específico, entre otros.

El ICEC-SP proporciona una puntuación global de cumplimiento y destacar áreas con mayor oportunidad vs esfuerzo (HH). Esto permite a la organización centrarse en los elementos comunes que tienen el mayor impacto en la seguridad de procesos y asignar recursos de manera más efectiva para mejorar la implementación de PSM.

El abordaje de está enfoque sé comparte mediante ejemplos reales, durante la presentación examinaremos cómo esta perspectiva puede conducir a mejoras significativas en la seguridad de procesos. Identificaremos los desafíos comunes a los que nos enfrentamos y las estrategias efectivas para superarlos. Enfatizaré que la seguridad de procesos no debe considerarse simplemente como una meta a alcanzar, sino como un viaje continuo que demanda un compromiso constante y una vigilancia rigurosa.

Desafíos típicos en la implementación de PSM:

- **Cambio Cultural:** Cambiar la cultura organizacional para priorizar la seguridad sobre otros objetivos.
- **Compromiso de la Alta Dirección:** Obtener el compromiso y el apoyo continuo de la alta dirección para la implementación de PSM.
- **Recursos Financieros y Humanos:** Garantizar la asignación adecuada de recursos financieros y humanos para la implementación y mantenimiento del PSM.
- **Identificación de Peligros y Evaluación de Riesgos:** Identificar todos los peligros potenciales y evaluar los riesgos asociados a los procesos.
- **Capacitación y Concientización:** Asegurar que todos los empleados estén capacitados y conscientes de los riesgos y las medidas de seguridad.
- **Gestión de Cambios:** Implementar procedimientos efectivos para gestionar cambios en procesos, equipos o sistemas.
- **Cumplimiento Normativo:** Cumplir con las regulaciones y estándares de seguridad puede ser un desafío, especialmente en entornos altamente regulados.
- **Prácticas de Mantenimiento, Confiabilidad e Integridad de Equipos e Instalaciones:** Asegurameinto de los equipos en condiciones óptimas y confiables.
- **Estrategias de Comunicación:** Desarrollar sistemas efectivos de comunicación interna y externa, y gestionar situaciones de crisis.
- **Evaluación de Desempeño y Mejora Continua:** Establecer indicadores clave de desempeño (KPI) y procesos para evaluar el PSM y mejorar continuamente la seguridad.

En síntesis, mi intervención se enfoca en compartir experiencias concretas, mejores prácticas y recomendaciones que pueden asistir a las organizaciones en el fortalecimiento de su cultura de seguridad de procesos y en el logro de un equilibrio efectivo entre la eficiencia y la seguridad. Al priorizar la adherencia por encima de la ruta típica, aseguramos que nuestros procesos sean seguros y confiables a largo plazo, sin comprometer su integridad ni su calidad durante el proceso.

Cómo Medir La Evolución Cultural De La Seguridad De Procesos. Casos De Éxito.

Martin Fernandez

La transformación de la Cultura de Seguridad de Procesos requiere una estrategia de trabajo en dos dimensiones principales. Por un lado, impulsar el Sentido de Vulnerabilidad en la Alta Gerencia, y por otro, impulsar la Disciplina Operacional en los frentes de trabajo y la toma de decisiones en función de riesgos, combatiendo la normalización de desvíos.

Sin embargo, para alimentar esta estrategia de transformación, es necesario desplegar una práctica consistente de medición y análisis de la cultura de seguridad de procesos, que aborde los ejes centrales de la gestión de riesgos e impactos, sus oportunidades de mejora y cómo estas evolucionan en el tiempo. Medir la cultura permite ajustar las tácticas y estrategias desplegadas para su evolución.

En este trabajo, completamos un proceso de cinco años (2019-2023) y dos ejercicios de medición cuantitativa que arroja resultados “en números” sobre un caso de evolución cultural. Detallar el proceso de análisis e interpretación de los resultados cualitativos de la cultura, a la vez que visualizarlos en un proceso evolutivo permitió a los líderes focalizar en aspectos críticos, y articular y desplegar las estrategias adecuadas, con el fin último de que el modelo cultural definido por la organización se vea reflejados en los comportamientos cotidianos.

A modo de nota metodológica, explicitamos que las dimensiones de cultura utilizadas en las encuestas implementadas por organizaciones que se presentan en el presente trabajo fueron elaboradas a partir de los core principles definidos en el primer elemento “Process Safety Culture”, del pilar “Commit to Process Safety”, en el marco del Process Safety Management System (Risk Based Process Safety, RBPS 2007)-, y a partir de la Escalera de Cultura (IOGP, práctica 435, 2011).

Como Migrar Hacia Un Proceso De Gestión De Cambio (MOC) Electrónico Con Éxito

Julio Mimbela

La Gestión de Cambio (MOC) es un proceso que tiene la intención de garantizar que cualquier cambio propuesto sea reconocido, revisado exhaustiva y sistemáticamente y que los impactos sean evaluados, comunicados a todas las partes afectadas, documentado, justificado y aprobados antes de su respectiva implementación y de esa manera eliminar la ocurrencia de eventos no deseados que conlleven a eventos catastróficos.

El Elemento de Gestión de Cambio formar parte del Sistema de Gestión de Seguridad de Procesos en base a la CCPS y OSHA respectivamente. Adicionalmente es uno de elementos del Sistema de Gestión de Procesos de la legislación peruana (Resolución de Consejo Directivo-203-2020-OS/CD)

Como parte del proceso de implementación del Sistema de Gestión de Procesos en PERU LNG, se implementó el proceso de Gestión de Cambio (MOC), en el cual cada una de las etapas del flujograma del proceso se manejaba de manera manual, con el inconveniente de no tener un control confiable (perdida de registro, demoras en el proceso de autorización entre otros) asociados al seguimiento de los pasos establecidos en el Procedimiento de Gestión de Cambio.

Esta implementación ha producido un gran impacto de la Operación de nuestras instalaciones permitiendo entre otros aspectos los siguiente:

- Mejor trazabilidad de los MOC´s presentados e implementados.
- Disminución en los costos de horas hombre (firmas de autorización de manera electrónica)
- Orden en el flujo del proceso de Gestión del Cambio

El principal objetivo de esta presentación es presentar una solución electrónica de éxito que permitió gestionar el proceso del ciclo de vida del proceso de MOC en la Planta de Procesamiento de Gas Natural de PERU LNG.

Gestión De Un Sistema Hips En Operación y Mantenimiento

Marcos Alamo Sr.

Operar y mantener el SIS reposa en 3 aspectos: La operación, que implica la interacción entre el personal y el SIS para que este funcione dentro de los límites y los intervalos de tiempo esperados; el mantenimiento, el cual conlleva a un trabajo diario sistemático para mantener el desempeño e integridad de lo diseñado y finalmente la gestión lo que incluye, registros, coordinación, planificación y competencia del personal. Dentro de los tipos de arreglos de funciones instrumentadas de seguridad donde podemos revisar estos tres aspectos tenemos al Sistema de protección de presión de alta integridad (HIPS), el cual se aplica para la protección específica contra escenarios de sobre presión evitando que esta se propague a puntos en donde no pueda ser soportada por la instalación o en donde otros dispositivos de alivio pudieran crear riesgos adicionales o ser imprácticos de ser aplicados como solución.

Por el lado normativo, el API 521, ASME sección VIII, API 14C e IEC 61511 refieren el uso de un sistema instrumentado de alta integridad como alternativa a los métodos tradicionales de alivio de presión que emplean dispositivos mecánicos. En la etapa de Operación y Mantenimiento, tenemos que asegurar que el *SIL requerido para nuestra HIPS se mantenga a lo largo de toda la vida operativa, además de que esta sea operada y mantenida asegurando la integridad de seguridad requerida.*, para ello, es necesario identificar entre otros los mecanismos de degradación del SIS los cuales pueden venir de dos fuentes principalmente, los operacionales y los de mantenimiento. Asimismo, no se debe perder de vista el art. 5.2.5.3 de la IEC 61511-1:2017 referido al seguimiento del desempeño.

En la etapa operativa también se deberá observar la importancia de las pruebas periódicas. Esta cláusula contiene un aspecto relevante a resaltar para ser observado tanto por el operador como por el mantenedor: Realizar todas las pruebas periódicas de nuestra SIF asociada a nuestra HIPS acorde a los periodos de tiempo especificado en la SRS, caso contrario, la probabilidad de falla de esta función irá incrementándose, reduciendo la disponibilidad de la seguridad asociada a esta HIPS. Adicionalmente, también le deberemos añadir el riesgo de tener un "proof test imperfecto"

Las actividades de Gestión asociadas en esta etapa deberán estar definidas, asignadas y alineadas según lo declarado en el Plan de Gestión de Seguridad funcional, así como todas las especificaciones del hardware y programa de aplicación en la SRS correspondiente. Consideraciones como tiempo máximo permitido, medidas compensatorias, actividades antes y después de remover un bypass deberán estar definidas en la SRS correspondiente

Un capítulo aparte es la necesidad de garantizar la competencia. La IEC 61511 lo define como: *"Las personas, departamentos u organismos implicados en el ciclo de vida de la seguridad de los SIS deben ser competentes para las actividades que son responsables",* lo que implica: comprender el propósito y los requisitos funcionales del SIS; comprender los peligros contra los que protege el SIS; tomar conciencia de las limitaciones operativas con las que operará el SIS; estar familiarizado con los procedimientos de prueba y mantenimiento relevantes, incluidos los procedimientos para el registro y la clasificación de fallas, así como estar familiarizado con los procedimientos de gestión de cambios (MOC).

Conclusiones

El Sistema de protección de presión de alta integridad (HIPS), permite atender un escenario particular de sobrepresión en donde los sistemas de alivio tradicionales son ineficientes. Asimismo, reduce la probabilidad de que varios dispositivos de alivio operen de manera simultánea, lo cual permite una reducción en el tamaño del sistema de disposición final, permite minimizar el uso de un diseño inherentemente seguro para escenarios de sobrepresión aguas abajo del sistema HIPS y también se convierten en una opción viable en donde las regulaciones medio ambientales restringen el venteo y quema de gas. Finalmente, se cuenta con información acerca de las consideraciones de su implementación en estándares internacionales reconocidos.

Si bien se trata de sistemas paquetizados con una arquitectura de instrumentación típica, debido a que los sistemas HIPS presentan restricciones de uso relacionados con su diseño, es necesario dimensionarlos según el equipo o sistema a proteger por ejemplo separadores de gas, equipos de proceso, ductos. De otro lado, para su implementación se requiere desarrollar el ciclo de vida definido en IEC 61511-1 lo que implica desde la implementación de un estudio de riesgo, el cumplimiento de procesos de verificación, evaluación y validación a lo largo de toda su vida útil, así como el desarrollo de un plan de gestión de seguridad funcional donde se designen roles y responsabilidades para cada fase, siendo la más larga la de Operación y mantenimiento.

¿Predecir El Pasado? Análisis De Consecuencias Durante La Investigación De Incidentes

Alejandro Jimenez

Muchas instalaciones industriales manejan y producen materiales inflamables, combustibles, y reactivos conllevando un riesgo inherente de la ocurrencia de incidentes de procesos durante sus operaciones. Esto es evidente al observar eventos catastróficos en las noticias como Flixborough, Bhopal o Piper Alpha. Sin embargo, no todos los incidentes son catastróficos, por lo que es importante la realización adecuada de una investigación de incidente como medio de prevenir dichos eventos. De acuerdo a la guía de investigación de incidentes de procesos del centro de seguridad química de los procesos (CCPS) el rol de una investigación de incidente incluye:

1. Prevenir eventos catastróficos a través de la evaluación de eventos menor severidad como eventos casi-accidentes (Near-Miss).
2. Prevenir desastres a través de la evaluación de eventos más serios y aprender de los mismos.

Existen diversas metodologías y herramientas usadas durante estas investigaciones. Una de las herramientas más usadas es el análisis de consecuencias como medio de entender en mayor detalle el evento. Estas herramientas permiten evaluar la magnitud de los eventos potenciales permitiendo una evaluación más detallada del evento. Sin embargo, estos modelos fueron desarrollados para ser utilizados en actividades de análisis de riesgo, por lo que su uso dentro de la investigación debe seguir una estrategia diferente.

En primer lugar, hay que entender que estos modelos tienden a poseer un carácter empírico, por lo que poseen una incertidumbre asociada. Por tal motivo, estos suelen tener un factor de seguridad asociados de forma que proporcionen valores relativamente conservadores respecto a un evento dado. Esta característica es totalmente aceptada dentro de un contexto de análisis de riesgo, pero genera un desafío importante durante la investigación de incidentes. De tal forma, para poder utilizar las herramientas de análisis de consecuencias efectivamente durante una investigación se debe entender y gestionar las incertidumbres presentes en estos modelos. Este entendimiento permitirá relacionar la predicción del modelo con el comportamiento de los materiales en el evento ocurrido.

Debido a estas incertidumbres, se debe identificar y categorizar adecuadamente la data relevante. Por un lado, se tiene la data observada el día del evento y por otro lado se tiene data empírica relacionados a las observaciones experimentales usadas para desarrollar las correlaciones. Durante el análisis se debe balancear adecuadamente estos datos con el fin de implementar adecuadamente el modelo que simule el evento ocurrido.

Finalmente, debido a la variabilidad de datos y la incertidumbre asociada con el uso de estos modelos, la implementación del análisis busca la presentar un rango de valores. Esta característica diverge del uso común de los modelos durante las actividades de evaluación de riesgo donde se tiene un valor determinado. De tal forma, que durante una investigación se implementa un análisis de sensibilidad ubicado alrededor del valor real generado por el evento.

Esta presentación y articulo describe la estrategia a seguir para adecuadamente usar los estudios de análisis de consecuencias durante una investigación de incidentes. Este trabajo mostrará inicialmente una explicación del concepto de análisis de consecuencias, describirá sus incertidumbres y la importancia de verificación y validación de estos, seguidamente mostrará la estrategia de implementación durante la investigación de incidentes y culminará con un caso de estudio donde se implementa un análisis de consecuencias durante una investigación de incidentes.

Implementación De Ventanas Operativas En Planta Deshidratadora LH3

Enrique Gonzalo Echave, Eduardo Roberto Ortiz, Maria Eugenia Perez Prieto

Las variables de los procesos industriales, y más específicamente los del **OIL&GAS**, deben ser observados y analizados constantemente por todas las estructuras de la compañía, desde un operador de planta, encargado de turno, ingenieros, supervisor o gerente de operación, ya sea para su optimización energética, confiabilidad del activo, seguridad de proceso u operar dentro de nuestros valores óptimos definidos.

Algo tan simple como suena, llevarlo a la práctica no es un trabajo sencillo de ejecutar, siendo el mayor desafío encontrar una herramienta que logre la simplicidad necesaria para que todos los actores interpreten objetivamente las mismas variables operativas de una misma forma y que sea la guía de operación para nuestros operadores diarios de la planta.

Con el uso de la aplicación PI PROCESS BOOK y PI VISION de la empresa OSISOFT se desarrolló un **tablero de visualización y seguimiento de VENTANAS OPERATIVAS** de planta deshidratadora Las Heras 03, perteneciente a la compañía YPF S.A ubicada en la cuenca del golfo San Jorge en la provincia de Santa Cruz, el cual nos permite el seguimiento de las variables críticas de nuestro proceso y en caso de tendencia de desvió el operador logre corregir con celeridad la causa originaría.

Gestión De Emergencias: Mega Sistema De Red Contra Incendios (RCI) En Plantas Químicas Industriales y Los Principales Desafíos De Seguridad De Procesos

Jaime Alberto Garrido Ruiz Sr., Roberto Guzmán Sr.

Introducción

Las plantas químicas industriales representan entornos críticos donde la seguridad y la gestión de emergencias son fundamentales. Entre las infraestructuras esenciales en estas instalaciones se encuentra el Megasistema de Red contra Incendios (RCI). En este paper, exploraremos los desafíos de seguridad de procesos asociados con grandes RCI en plantas químicas industriales y las estrategias para abordarlos.

Megasistema de Red contra Incendios (RCI)

Un Megasistema de Red contra Incendios (RCI) es una red integral de protección contra incendios que abarca toda una planta química industrial. Incorpora tuberías, bombas, hidrantes o monitores, aspersores y sistemas de detección y control de incendios. Los RCI son cruciales para la prevención y control de incendios en entornos donde la manipulación de sustancias químicas peligrosas presenta riesgos significativos.

Desafíos de Seguridad de Procesos en Grandes RCI

- **Diseño y Mantenimiento Complejo**: Los grandes RCI son sistemas altamente complejos que requieren diseños precisos y mantenimiento riguroso. La American Society of Mechanical Engineers (ASME) enfatiza la importancia del diseño de sistemas de RCI en su Código ASME A17.1, que establece normas para estos sistemas. El mantenimiento adecuado se respalda con cifras que indican que el 80% de los fallos en sistemas contra incendios ocurren debido a un mantenimiento insuficiente.
- **Suministro y Presión de Agua**: Asegurar un suministro de agua constante y la presión necesaria para combatir incendios puede ser un desafío, particularmente en ubicaciones con limitaciones de agua. Según National Fire Protection Association (NFPA), las estadísticas muestran que en el 44% de los incendios industriales, los sistemas de RCI no funcionan debido a problemas de presión de agua inadecuada.
- **Detección y Control Avanzado**: Los sistemas de detección y control de incendios deben ser avanzados y altamente sensibles. NFPA señala que en el 60% de los casos de incendio en plantas químicas, la detección temprana es clave para prevenir la propagación del fuego.
- **Compatibilidad de Sustancias**: El Centro de Investigación y Desarrollo de Protección contra Incendios (Fire Protection Research Foundation) advierte que la incompatibilidad química con materiales y agentes extintores puede desencadenar reacciones peligrosas en caso de incendio, lo que agrega un nivel adicional de riesgo.
- **Entrenamiento y Respuesta Efectiva**: Según un informe de National Safety Council (NSC), la capacitación inadecuada contribuye al 40% de los incendios industriales. La respuesta efectiva a un incendio en una planta química depende en gran medida de la formación del personal y su capacidad para utilizar equipos de protección y control de incendios de manera adecuada.

Estrategias para Abordar los Desafíos de Seguridad de Procesos en Grandes RCI

- **Mantenimiento Preventivo**: Según el National Institute for Occupational Safety and Health (NIOSH), los programas de mantenimiento preventivo rigurosos reducen la probabilidad de fallos del sistema RCI y son críticos para mantener la integridad del sistema.
- **Simulacros y Entrenamiento Continuo**: NFPA recomienda simulacros regulares para el personal de la planta química. Estos simulacros pueden mejorar la respuesta de emergencia y aumentar la familiaridad del personal con los procedimientos de seguridad.
- **Monitoreo Avanzado**: Utilizar sistemas de monitoreo y control avanzados respaldados por la American Industrial Hygiene Association (AIHA) para garantizar una respuesta más rápida y precisa en situaciones de incendio.
- **Actualización de Diseño**: La AIHA también enfatiza la importancia de revisar y actualizar el diseño del sistema RCI para abordar los cambios en la planta y las sustancias químicas manejadas.
- **Planificación de Contingencia**: El Emergency Management Institute (EMI) de FEMA proporciona pautas específicas para la planificación de contingencia en situaciones de incendio, lo que incluye la gestión de sustancias peligrosas y la evacuación.

Caso Práctico: RCI SQM Yodo Nueva Victoria, Chile

- RCI con más de 7 Km de extensión se presenta con video introductorio y despliegue de elementos de gestión de seguridad de procesos.

Conclusión

La gestión de emergencias en plantas químicas industriales, con un enfoque en los Mega sistemas de Red contra Incendios (RCI), es un desafío crítico respaldado por datos y cifras que subrayan la importancia de abordar los riesgos. Los desafíos de seguridad de procesos en grandes RCI, desde el diseño y mantenimiento complejo hasta la respuesta efectiva, requieren un enfoque integral y un compromiso continuo. La gestión efectiva de los Mega sistemas de RCI es vital para proteger vidas, instalaciones y el medio ambiente en estos entornos de alto riesgo en las plantas químicas industriales.

Cómo Mejorar La Eficiencia De Los PHA, Maximizando El Brainstorming e Identificando Escenarios De Mayor Relevancia

Matias Massello

Este paper incluye las lecciones aprendidas en más de 8000 evaluaciones cualitativas de peligros realizadas por los autores en los últimos 40 años. Condensa la experiencia de dichas evaluaciones de peligros (las cuales incluyen: proyectos, MOCs, unidades/plantas en operación y revalidaciones) con el fin de proporcionar un listado de las actividades que optimizan los recursos necesarios para realizar los PHA y maximizar el brainstorming sin dejar de ser exhaustivos.

En este paper se incluyen las respuestas a preguntas como:

- ¿Es necesario tener una persona dedicada exclusivamente a documentar la reunión (Escriba)?,
- ¿Es necesario utilizar un software específico?,
- ¿Qué conviene proyectar durante las reuniones?,
- ¿Cuándo conviene realizar la demarcación de Nodos/Sistemas?,
- ¿Es recomendable utilizar siempre la misma metodología?,
- ¿Qué estilo de documentación utilizar?,
- ¿Qué reglas debe seguir el Líder agilizar la reunión sin impactar en el brainstorming?,
- ¿Qué actividades impactan negativamente en la velocidad de reunión y en el brainstorming?,
- entre otras.

Defensa Inquebrantable: Desentrañando La Efectividad De Barreras y Criterios ALARP En La Industria De Los Hidrocarburos En El Perú – Comité 172 Inacal

Cesar Puma Medina Sr., Margot Lopez Garcia, Oswaldo Mata Moreno

Defensa Inquebrantable: Desentrañando la Efectividad de Barreras y Criterios ALARP en la Industria de los Hidrocarburos en el Perú – Comité 172 INACAL

La industria química y del Oil&Gas es conocida por sus riesgos inherentes debido a la manipulación y producción de sustancias químicas peligrosas. En este contexto, la implementación de barreras de seguridad y la aplicación de criterios ALARP (As Low As Reasonably Practicable, tan bajo como sea razonablemente posible) son elementos esenciales para garantizar un entorno de trabajo seguro y prevenir incidentes graves. Este explora resumen la importancia de la efectividad de las barreras de seguridad y la aplicación de criterios ALARP adecuados al entorno operativo local en la industria química, y cómo estas estrategias contribuyen a un entorno de trabajo más seguro y sostenible. En tal sentido se formó en el Perú el Comité 172, encargado de normalizar y estructurar los criterios de PSM para la industria nacional.

Importancia de la Efectividad de Barreras de Seguridad en la Industria Química

La seguridad en la industria química es de suma importancia, ya que un solo incidente puede tener consecuencias devastadoras en términos de lesiones, daños ambientales y financieros. Las barreras de seguridad son las medidas y sistemas diseñados para prevenir incidentes y mitigar sus impactos. La efectividad de estas barreras es crucial para proteger a los trabajadores, el medio ambiente y los activos de la empresa.

Importancia de establecer Criterios ALARP en la Industria Química del Perú

Los criterios ALARP son una herramienta de gestión de riesgos que se centra en reducir los riesgos a un nivel 'tan bajo como sea razonablemente posible'. Esto significa que las medidas de seguridad deben ser proporcionales al riesgo y económicamente viables. Los criterios ALARP son esenciales para equilibrar la seguridad con la eficiencia operativa y garantizar que los recursos se utilicen de manera eficaz.

Sinergia entre Barreras de Seguridad y Criterios ALARP

La sinergia entre las barreras de seguridad y los criterios ALARP es fundamental. Las barreras de seguridad se convierten en las herramientas prácticas para implementar los criterios ALARP. Al evaluar los riesgos y aplicar medidas de seguridad, se busca lograr el equilibrio entre la seguridad y la viabilidad económica. Esto se traduce en la implementación de barreras de seguridad efectivas que sean proporcionales al riesgo.

Efectividad de las Barreras de Seguridad y su Medición

La efectividad de las barreras de seguridad se mide en términos de su capacidad para prevenir incidentes y mitigar sus impactos. Algunos de los indicadores claves de la efectividad de las barreras de seguridad incluyen:

- **Prevención de Incidentes Graves:** Las barreras de seguridad deben prevenir incidentes graves que podrían resultar en daños humanos, daños ambientales significativos o pérdidas económicas importantes.

- **Tiempo de Respuesta Eficaz:** Las barreras de seguridad deben permitir una respuesta rápida y eficaz en caso de incidentes menores, minimizando la propagación de los mismos.
- **Cumplimiento Regulatorio:** Las barreras de seguridad deben cumplir con las regulaciones y estándares aplicables en la industria química.
- **Optimización de recursos:** Las barreras de seguridad deben ser eficientes en términos de recursos utilizados, evitando gastos innecesarios sin comprometer la seguridad.

Implementación de Barreras de Seguridad y Criterios ALARP

La implementación exitosa de barreras de seguridad y criterios ALARP implica una serie de estrategias:

- **Evaluación de Riesgos Integral:** Una evaluación exhaustiva de riesgos debe identificar y cuantificar los peligros potenciales. Esto sirve como base para la implementación de medidas de seguridad efectivas.
- **Selección de Barreras Adecuadas:** Se deben seleccionar las barreras más adecuadas para abordar los riesgos identificados. Esto puede incluir sistemas de alarma, procedimientos de respuesta a incidentes, sistemas de cierre de emergencia y dispositivos de seguridad.
- **Capacitación del Personal:** El personal debe recibir capacitación adecuada en la operación de barreras de seguridad y la comprensión de los criterios ALARP.
- **Auditorías y Evaluaciones Periódicas:** La efectividad de las barreras de seguridad debe ser monitoreada y evaluada regularmente a través de auditorías y revisiones.

Estudios de Casos y Resultados Prácticos

Los estudios de casos en la industria química ilustran cómo la efectividad de las barreras de seguridad y la aplicación de criterios ALARP han prevenido incidentes significativos. Estos ejemplos prácticos ayudan a destacar la importancia de estos a través del enfoque desarrollado por el Instituto Nacional de Calidad INACAL-Comité 172 del Perú.

Responsabilidad Ética y Sostenibilidad

La implementación de barreras de seguridad y criterios ALARP en la industria química no solo se trata de cumplir con regulaciones y normativas, sino también de asumir la responsabilidad ética de proteger a las personas y el medio ambiente. Además, contribuye a la sostenibilidad al reducir los riesgos de incidentes que podrían tener un impacto ambiental negativo.

Conclusiones y el Futuro de la Industria Química

La efectividad de las barreras de seguridad y la aplicación de criterios ALARP son elementos cruciales para garantizar un entorno de trabajo seguro y sostenible en la industria química. Al equilibrar la seguridad con la eficiencia, las organizaciones pueden prevenir incidentes graves y garantizar la protección de las personas y el entorno. Esto sienta las bases para un futuro más seguro y ético en la producción química.

164b Fame or Blame do not heal the pain of losing loved ones

Kumar Israni

When we hear the terms Process Safety Management (PSM), the human mind, directly associates this with the processing and manufacturing industries. How many times, have family dinner conversations, always mocked at the person that canvasses people or process safety? "Leave this at work", is a common response in the event of such discussions. But are people and process safety just limited to the work environment?

As process safety professionals, we always talk about planning and management of disasters and other unexpected incidents that carry a potential to harm human lives and the environment, the Emergency Response Plans (ERP) being an effective tool to aid at these times. Infact, these ERPs are critical even in a non-PSM environment. The recent Astroworld stampede in Houston TX, while not a PSM type of scenario, still warranted the need for an effective ERP.

The author's 18-year-old son attended the Astroworld event of 2021 in Houston, TX and witnessed the disaster first hand, along with the lack of appropriate rescue measures. Having been in the stampede, and luckily escaping out of it along with a rescue of a friend, became his heroic accomplishment for the day, however his first reactions were the lack of communication and inadequate measures for rescue that were witnessed by him and his friends. He came home, that night, trembling; waking up his father letting him know that he survived a disaster and thanking his parent for the imbibed safety coaching. The next morning news reports indicated the loss of 10 lives at this event.

Fast forward, year 2023 the Houston Police Department has released its formal report. Also the author's older son who is now 23 years old, attended a concert by the same performing artist in Miami, FL, which occurred approximately two years after the disaster event of 2021. This event was a successful event with no loss of lives.

This presentation will discuss the process safety aspects that include findings from the disaster of 2021 and highlight the modifications to Emergency Response and other safety aspects seen in 2023. Also, it will include the impact of incidents on human lives, using this family's current appreciation on the value of human lives and the importance of ERP and PSM even in a non-PSM or day-to-day environment be it the workplace or public place or even the home. People and Process safety discussions are willingly embraced by the family now, not a topic of any mockery anymore.

146a Common Mistakes with the Application of a Risk Matrix

Timothy Stirrup, Kelsey Forde

This education session is based on the risk assessment principles and methodologies from publications of the Center for Chemical Process Safety (CCPS), American Industrial Hygiene Association (AIHA), International System Safety Society (ISSS), American Society of Safety Professionals (ASSP), and Alliance of Hazardous Materials Professionals (AHMP). The presentation will discuss lessons learned from the use of qualitative, semiqualitative, and qualitative risk matrices. Risk management guides and references from these organizations as well as additional standards (e.g., ANSI, SEMI) provide the hazards analysis (HA) practitioner with guidance on the use of risk matrices to support the identification of risk. Risk matrices are a tool in the HA practitioner's toolbox to help make risk-based decisions to assign resources. Proper use of the risk matrix is a powerful tool; whereas, misapplication of the risk matrix may yield a hollow promise of safety and waste resources. The information and lessons learned provided will aid the HA practitioner with addressing unmitigated risk, mitigated risk, and adequacy of safeguards.

Learning Outcomes

Upon completion, the participant will be able to:

• Use qualitative risk matrix development principles.

• Identify common mistakes in risk matrix usage.

• Calculate consequence & frequency derivations.

• Utilize preventive & mitigative control applications.

146b Major Chemical Safety Incidents – a Review

Dana Garber

This session aims to review incident investigations from The U.S. Chemical Safety and Hazard Investigation Board (CSB) with a focus on gathering data on common causes & failure modes, Severity data, as well as any links to the facility's EPA compliance and major incidents.

The CSB is an independent federal agency charged with investigating industrial chemical accidents. The CSB conducts root cause investigations of chemical accidents at fixed industrial facilities. Root causes are usually deficiencies in safety management systems but can be any factor that would have prevented the accident if that factor had not occurred. Other incident causes often involve equipment failures, human errors, unforeseen chemical reactions, or other hazards.

In this session, attendees will:

- Receive an overview of many major industrial chemical incidents over the last 10 years within the US.
- Identify commonalities between the incidents.
- Understand how failures within a safety management system can be key factors that lead to high hazard facility incidents.

66a Reactive Chemicals Hazard Introduction: An Industrial Perspective

Katie Mulligan

Chemical reactivity hazards present complex challenges in industrial settings, encompassing various types of reactions and chemical substances. This presentation delves into the diverse array of reactivity hazards, including the risks associated with monomers, oxidizers, runaway reactions, autoignition phenomena, organic peroxides, highly energetic materials, and pyrophoric chemicals. Through an exploration of the characteristics and behaviors of these substances, attendees will gain insights into the mechanisms driving hazardous reactions and the potential consequences of uncontrolled reactivity. By fostering a deeper understanding of chemical reactivity hazards and equipping attendees with identification approaches, this presentation aims to enhance safety awareness and promote a culture of prevention.

66b Hazard Analysis and Risk Assessment Methods: How to Choose…

Michael Moosemiller, Nathan Knochel

Hazard Analysis and Risk Assessments are commonly used by organizations to ensure that people, the environment, and business continuity are protected. Various approaches and methodologies can be used to perform these assessments. Some approaches are very simple, require minimal information, time and subsequently have a low resource impact on an organization, but as a result may provide only broad input to decision-making. Use of more detailed approaches, such as a detailed dispersion models, development of fault trees, or performing a Quantitative Risk Assessment (QRA) often yield a less conservative and more resolute estimation of consequence and risk, and can provide precise engineering guidance such as where to use emergency isolation, or whether a proposed mitigation measure is worth the investment. The use of more detailed approaches significantly increases the resource impact on an organization including the use of complex tools / software and time investment in performing these detailed evaluations. Understanding when and how to use the right approach, at the right time, is key. This brief tutorial will provide an overview of various approaches to evaluate the consequences of identified hazards and assess the associated risk.

Our First Taste of Process Safety Culture, and How It Impacted Our Lifestyles.

Radhika Bompelly, Kumar Israni

Our journey in process safety began several years back. However, out taste of the American culture began prior to that. Having had overseas education in engineering, before we moved to the United States at different times, were a huge change in culture, and mindset for us. Differences between a developed nation and a developing country, can be huge; was our big learning in that process. This was observed in every aspect of engineering – concept, application, and implementation.

Early years of engineering, while limited to the process world, brought constant comparative apprehensions in how engineering was practiced between Asia and North America. A turning point emerged as we entered the world of process safety. This was an evolutionary journey for us. Understanding the expectations of OSHA PSM and EPA RMP regulations, and how they were implemented were a huge eye-opener.

As we got deeper into this process safety and risk management journey, regular exchanges of ideas and observations, made us conclude that while there are regulatory differences in various regions of the world, importance of process safety culture is literally embracing a safety mindset with the constant reminder that safety must be practiced 24/7 and there is no holiday for safety - be in people safety or process safety.

The authors share their experiences of moving to the United States at different times in their careers in Process Safety, while bringing out the common learnings of human performance along with the importance of the value of human lives. There was a common understanding in these core values in various regions of the world, however was practiced and implemented in different ways, resulting in good and bad outcomes.

Maintain Curiosity - Process Safety Culture at Gibson Energy

Connor Murray

A campaign with the slogan "Maintain Curiosity" was recently initiated at Gibson Energy to drive a culture change towards a risk-based process safety culture and to increase process safety management competency and performance among front-line personnel. The slogan emphasizes the importance of sustaining curiosity towards process safety while conveying that front-line employees baseline knowledge, skills and abilities is essential for the campaign's success. The campaign involves a three-phased rollout which will include process safety fundamentals, identifying process safety hazards and risks and the effective management of these risks.

22a Frontline DEI: How to Design and Integrate Inclusion into Team Processes

Jody Jahn

Organizations across sectors are placing strategic priority on diversity, equity, and inclusion (DEI) efforts. However, several case studies show that current approaches—particularly, efforts to change the individual organizational member—are not working as well as they could. Many DEI consulting firms aim to change culture by changing individual members by, for example, putting them through implicit bias training, and engaging in group exercises that aim to build self-awareness about different perspectives. While important efforts, individual-level development can only impact group culture indirectly, at best. Instead, to shape team safety culture, leaders should intervene directly on group-level behaviors. Group-level behaviors are distinct from individual-level behavior in that group behavior is defined as the collective action of multiple interdependent actors. We see group-level behavior unfold through team routines and practices, and in the nature of the interaction patterns that team members regularly engage in with each other (e.g., as informal mentors/mentees, as competitors, as equals/peers). This talk first defines psychological safety, diversity, equity, and inclusion, then operationalizes those terms according to what group actions you would observe if you were seeing these concepts working well versus working poorly within a group. I will then walk through an example of how to intentionally re-design a team practice to incorporate inclusion and psychological safety into the regular workflow of the group. This approach is different from common DEI approaches that rely on organization members to gain personal awareness about implicit bias and difference. Instead, a Frontline DEI™ framework enables a leader to design an inclusive safety culture by making small changes to existing group routines and practices, and intentionally designing new ones.

162a Implementing Simultaneous Operations at Superior Refinery

Brian Steinke, Ola Olamilehin

After a major fire incident at the Superior Refinery Wisconsin in 2018, a major rebuild project was executed to restore the facility with expanded processing capacity of 49,000 barrels per day (bbls/d) and a full slate of products, including asphalt, diesel, gasoline and heavy fuel oils. A re-start plan that addressed processing unit startups and simultaneous construction and commissioning work led to the implementation of a Simultaneous Operations (SIMOPS) Process. The SIMOPS process used an activity matrix tool for decision making. The SIMOPS activity matrix was developed with consideration of operational, construction, and commissioning activities. Each activity scenario on the matrix was than assessed and categorized as acceptable, not acceptable, or restricted. A SIMOPS (PIC) or Person-in-Charge was identified to implement the SIMOPS process during the construction, commissioning, and process unit startups. The implementation of the SIMOPS process allowed for work activities to safely continue while starting up process units, thereby accelerating the refinery re-start, leading to early production of asphalt and diesel products.

When Going Beyond Empirical Is Needed, a Real Life Example.

Vinicius Simoes, Drew Botwinick

Most facility siting analyses start with a basic, worst-case empirical consequence analysis. And as it turns out, many of those studies quickly demonstrate that the impact to buildings exceeds the acceptable criteria and require further analysis. The next level of analyses includes various techniques to rebalance conservatism and realism including quantitative risk-based studies that produce exceedance curves, etc. However, because the basis of these studies is generally trying to reach the acceptable criteria from the worst-case direction while minimizing the changes to the consequence modelling effort itself—it is not uncommon for these studies to result in design loads that require expensive retrofits to buildings. As a final step, some of these studies will turn to computational fluid dynamics (CFD) to justify avoiding these expensive retrofits. By following this process, the cost of the facility siting study is maximized and there is a higher chance that expensive building retrofits that don't actually pass a cost-benefit analysis might be forced to move forward.

This paper will demonstrate some of the benefits of using more realistic CFD modelling earlier in the process to more accurately reflect the real extent of potential building damage. When using the right mix of empirical modelling and CFD up-front as part of a risk-based analysis, facility siting studies can more realistically reflect the potential harm and provide greater cost-benefit—i.e., reduce the true overall cost of the study while also being more realistic about when it may be required to retrofit buildings to meet certain design accidental loads. The paper will focus on a real-life example of this application.

105a Thermal Protection Assessment on a Railcar Transporting Hazardous Material

Sebastian Chialvo, Alejandro Jimenez

This presentation describes a Fire Hazard Analysis (FHA) carried out in conjunction with the US Federal Railroad Administration (FRA) to evaluate the performance of a railcar's thermal protection system in cases of pool fires that affect the rail cars when they are located in the loading and unloading terminals of chemical and oil facilities. In this analysis, DOT-117 type cars were considered, which transport low-volatile flammable materials such as ethanol or crude oil. We assess the thermal protection system using a combination of computational fluid dynamics (CFD) modeling, thermodynamics modeling, and fire experiments. In addition to presenting the results of the assessment, this talk also discusses the importance of understanding the limitations imposed by one's choice of inputs and assumptions in the assessment.

105c Quantitative Risk Analysis Applied to Transportation of Acute Toxic Chemicals

Rafael Callejas-Tovar

Transportation of acute toxic chemicals is a key activity in the supply chain, which has been increasing due to market globalization. Routes might include railcar, truck, ship, crane movements, etc. and might pass through locations with high population, which may pose significant risks as an accidental release can potentially expose a large public population. This presentation will show the application of quantitative risk analysis (QRA) to determine the risk profile along the length of the route and help to identify potential high-risk locations and operations. In addition to discussing the consequence and frequency analysis methodology, refinements of the model using Failure Mode and Effects Analysis (FMEA), Finite element analysis (FEA), and other methods will also be presented. It will be demonstrated using case studies how the QRA results can be used for decision-making and developing comprehensive risk mitigation strategies.

AUTHOR INDEX